STATE OF THE SPACE INDUSTRIAL BASE

THREATS, CHALLENGES AND ACTIONS

T. COOLEY, E. FELT, AND J. BUTOW

NIMBLE BOOKS LLC: THE AI LAB FOR BOOK-LOVERS
~ FRED ZIMMERMAN, EDITOR ~

Humans and AI making books richer, more diverse, and more surprising.

Publishing Information

(c) 2023 Nimble Books LLC
ISBN: 978-1-60888-261-8

AI-generated Keyword Phrases

U.S. space industrial base; space dominance; national economic power; political power; military power; external threats; adversaries; China; dominant space power; internal challenges; industrial base; international adversaries; speed and innovation; upgrading methodologies; flexible markets; government procurement; licensing processes; unified national strategy; strong space industrial base.

Cover Art

The AI-generated prompt was: *Create a black and white illustration that conveys a somber and serious mood. The drawing should depict the vastness of outer space, with the earth and moon in the distance, symbolizing the importance of space to the nation. Show an astronaut floating in the foreground, seemingly alone and isolated, representing the challenges faced by the U.S. in maintaining its space dominance. Use strong contrast and shading to depict the threats posed by adversarial countries like China, with dark ominous clouds or symbols of competition looming in the background. The illustration should evoke a sense of urgency and determination, with the astronaut's posture and expression reflecting the need for strategies that will secure a strong and innovative space industrial base.*

FRONT MATTER

ABSTRACTS

TL;DR (ONE WORD)

Space.

EXPLAIN IT TO ME LIKE I'M FIVE YEARS OLD

This document is like a special report that talks about how important space is to our country. It says that space is important for our economy, politics, and military power. It also talks about how other countries, like China, are getting really good at space stuff and might be better than us soon. The document also talks about the challenges we face in keeping up with them and coming up with new and better ideas. It gives some ideas on how we can do better, like improving how we do things

TL;DR (VANILLA)

This white paper discusses the threats and challenges to maintaining the U.S. space industrial base and dominance in space. It highlights China as a major adversary and recommends strategies for staying ahead, such as upgrading methodologies and government processes. A unified national strategy is emphasized.

SCIENTIFIC STYLE

This white paper presents an analysis of the threats, challenges, and necessary actions to maintain the United States' space industrial base and space dominance. It emphasizes the growing significance of space in the context of national economic, political, and military power. The paper highlights the external threats posed by adversaries, specifically China, who are actively competing with and surpassing the United States as a dominant space power. Additionally, it addresses the internal challenges of establishing an industrial base that exceeds international adversaries in terms of speed and innovation. The paper recommends various strategies, including upgrading methodologies, developing flexible markets, and making changes to government procurement and licensing processes. It concludes by emphasizing the imperative need for a unified and comprehensive national strategy to construct and rejuvenate a robust space industrial base.

ACTION ITEMS

Invest in research and development to enhance technological capabilities and maintain a competitive edge in space.

Strengthen partnerships with allies and international organizations to promote cooperation and information sharing in space activities.

Increase funding and support for STEM education and workforce development to ensure a skilled and diverse workforce in the space industry.

Enhance cybersecurity measures to protect space assets and infrastructure from cyber threats.

Streamline government regulations and processes to facilitate innovation and commercialization in the space industry.

Encourage collaboration between government, industry, and academia to foster innovation and accelerate technology development.

Invest in infrastructure and facilities to support future space activities.

VIEWPOINTS

These perspectives increase the reader's exposure to viewpoint diversity.

GROUNDS FOR DISSENT

A member of the organization responsible for this document might have principled, substantive reasons to dissent from this report. Some potential reasons could include:

Overemphasis on military dominance: The document appears to prioritize maintaining space dominance from a military standpoint. A dissenting member may argue that this approach perpetuates an arms race mentality and fails to acknowledge the potential for peaceful cooperation and collaboration in space. They may advocate for a more balanced approach that promotes international cooperation and diplomacy rather than focusing solely on military superiority.

Neglecting environmental concerns: The document primarily focuses on economic, political, and military aspects of space, but it may fail to address potential environmental impacts. A dissenting member might argue that neglecting environmental considerations is not only ethically problematic but also detrimental to long-term sustainability. They may advocate for incorporating environmentally conscious practices into the strategies and actions recommended in the report.

Lack of international perspective: While the document acknowledges the competition with China, it may give insufficient attention to international cooperation and the importance of diplomacy. A dissenting member might argue that a unilateral approach focused on outpacing adversaries can hinder diplomatic efforts and potentially escalate tensions. They may advocate for greater emphasis on international partnerships and collaboration to address global challenges in space.

Inadequate consideration of societal implications: The document primarily focuses on economic and military benefits of the space industrial base, but it may overlook the potential societal implications. A dissenting member might argue that the report should address issues such as equity, accessibility, and the ethical use of space technologies. They may advocate for strategies that ensure the benefits of the space industrial base reach all segments of society and mitigate any negative consequences.

Lack of transparency and public input: The document, being a white paper, might not have involved sufficient public consultation or input from stakeholders outside the organization responsible for it. A dissenting member might argue that decisions about the future of the space industrial base should involve a broader range of perspectives, including those from academia, civil society organizations, and the public. They may advocate for a more inclusive and transparent approach to decision-making processes.

RED TEAM CRITIQUE

The document provides a comprehensive overview of the threats, challenges, and actions needed to maintain the U.S. space industrial base and space dominance. However, there are several areas where the document could be strengthened:

Lack of specific evidence: While the document highlights the increasing importance of space to national power and the aggressive actions of adversaries such as China, it fails to provide specific evidence or references to support these claims. Including concrete examples and data would strengthen the document's credibility and make its arguments more convincing.

Limited focus on technological innovation: The document briefly mentions the need for the U.S. industrial base to outpace international adversaries in speed and innovation. However, it does not delve deeply into strategies for fostering technological innovation. Identifying specific areas of technological advantage and proposing initiatives to maintain and enhance this advantage would add depth to the document.

Insufficient analysis of flexible markets: The recommendation to develop flexible markets is mentioned but lacks a thorough analysis. It would be beneficial to provide detailed insights into how flexible markets can be created and sustained, including potential challenges and considerations. Additionally, exploring the impact of market dynamics and competition on the space industrial base would further strengthen the document.

Limited discussion on procurement and licensing processes: The document suggests making changes to government procurement and licensing processes but lacks specific proposals or strategies for improvement. Providing specific recommendations, such as streamlining bureaucratic procedures or incentivizing private sector involvement, would enhance the document's practicality.

Lack of emphasis on international cooperation: While the document focuses on maintaining U.S. space dominance, it does not adequately discuss the importance of international cooperation in achieving this goal. Highlighting opportunities for collaboration with allies and partners, as well as addressing potential barriers to cooperation, would provide a more nuanced perspective.

Incomplete analysis of internal challenges: The document briefly mentions internal challenges related to developing an industrial base that outpaces adversaries, but it does not explore these challenges in detail. Identifying specific internal barriers and proposing strategies to overcome them would enhance the document's thoroughness.

Limited discussion on cybersecurity: Given the increasing reliance on space assets and the potential vulnerabilities to cyberattacks, it is surprising that the document does not dedicate more attention to cybersecurity. Considering the importance of protecting space assets from cyber threats, a more robust analysis of cybersecurity measures should be included.

In summary, while the document provides a solid foundation for understanding the threats and challenges facing the U.S. space industrial base, it could benefit from more specific evidence, deeper analysis of key areas, and concrete recommendations for improvement. Additionally, addressing international cooperation, cybersecurity, and internal challenges in greater detail would enhance the document's overall effectiveness.

MAGA Perspective

This document is just another example of the globalist agenda at work. The idea that we need to focus on maintaining our space dominance is just a distraction from the real issues facing our country. Instead of wasting time and resources on space, we should be focused on putting America first and addressing the problems right here at home.

It's laughable to think that China is somehow a threat to our space dominance. We are the United States of America, the most powerful country in the world. China can't compete with us on any level, let alone in space. This is just fearmongering designed to keep us in a perpetual state of war.

Furthermore, why should we be investing in the space industrial base when there are so many other pressing needs in our country? Our infrastructure is crumbling, our healthcare system is broken, and millions of Americans are struggling to make ends meet. We should be using our resources to fix these problems, not sending them off into space.

The strategies recommended in this document, like upgrading methodologies and developing flexible markets, are just empty buzzwords that do nothing to address the real issues. We need real solutions, not vague recommendations that sound good on paper but have no substance behind them.

Lastly, the idea of a unified national strategy for the space industrial base is just another way for the government to exert control over the private sector. We don't need the government interfering in our businesses and stifling innovation. We need to let the free market work and allow American entrepreneurs to thrive without unnecessary government interference.

PAGE-BY-PAGE SUMMARIES

0   *A workshop discussing the threats and challenges to the US space industrial base and space dominance, led by Dr. Thomas Cooley, Colonel Eric Felt, and Colonel Steven J Butow from the Air Force Research Laboratory and Defense Innovation Unit.*

1   *The page provides an overview of the Air Force Research Laboratory's mission in developing warfighting technologies for air, space, and cyberspace forces. It also discusses the Defense Innovation Unit's role in accelerating commercial innovation for national security, particularly in the space sector.*

2   *This white paper addresses the challenges and threats to the U.S. space industrial base and emphasizes the need for a coordinated national space strategy. It highlights the importance of a vibrant and competitive space industry to maintain national power and outlines potential strategies to address these challenges.*

3   *China poses significant threats to the US space industrial base through theft of intellectual property, integration of state-owned corporations with startups, penetration of American companies, investment in the US space industry, control of supply chains, and predatory pricing. Urgent action is needed to maintain US dominance.*

4   *The page discusses the threats and challenges facing the space industrial base, including state-sponsored technology surveillance and economic aggression by China. It emphasizes the need for a unified national strategy to develop and maintain a strong space industrial base.*

5   *The page discusses the need for a resilient space industrial base in the United States and the challenges and threats it faces. It emphasizes the importance of domestic manufacturing and defense supply chains and suggests potential strategies to address these issues.*

6   *The page discusses the growing role of space in delivering unique capabilities and advantages. It highlights the importance of space systems for information gathering, precision position, navigation and timing, and broadband communications. It also emphasizes the need for proactive measures to disrupt China's plan to become the dominant space power.*

7   *China's dominance in germanium mining and production poses a threat to the US space industrial base, as it controls the raw materials necessary for manufacturing satellite solar panels. The increased costs and limited availability of refined germanium metal give China influence over US producers and the availability of radiation hardened solar cells.*

8   *The page discusses the growing space ecosystem and the need for a strong space industrial base to support commercial, civil, and military applications. It also mentions the expansion of capabilities beyond low Earth orbit and the establishment of a permanent U.S. presence on the Moon and beyond. Collaboration with national commercial capabilities is crucial for a sustainable space economy.*

9   *The US space industrial base for high-efficiency solar cell production is significantly behind China, which has a much larger installed base of Metal Organic Chemical Vapor Deposition (MOCVD) reactors. The US will need to increase its production capacity to meet the growing demand for space solar power.*

10   *A diverse and robust national and international space industrial base is crucial for economic, political, and military superiority in space. A whole-of-government strategy and tailored space industrial policy are needed to address challenges and promote competition, innovation, and investment in the space sector. Over-reliance on government as the primary consumer hinders the growth of the space industrial base.*

11   *The page discusses the need for government acquisition approaches to adapt to the commercially-oriented space industrial base, address adversarial activity, and focus on reducing launch costs to maintain a competitive market.*

12   *China's state-owned defense contractor, CASIC, is offering its Kuaizhou-11 launch system at a highly discounted rate, threatening the viability of the global launch industry, including the US. This predatory pricing tactic has been successful for Chinese companies in other industries, such as DJI dominating the drone market.*

13   *The page discusses the shift from GEO to LEO satellites and the potential growth of the space industrial base. It also highlights China's strategy to become a dominant space power economically, politically, and militarily.*

14   *China's strategy to dominate the global space industrial base includes developing their own capabilities, stealing intellectual property, and controlling the supply chain. Counterintelligence efforts are needed to block adversary funding of US space technology innovators. Legal changes are necessary to protect emerging technologies. The dominance of Huawei in 5G communications exemplifies the challenges posed by China's space strategy.*

15   *China's dominance in the space market poses a threat to the US. A whole-of-government strategy is needed to address this and develop a tailored space industrial policy that combines government, industry, and academia. The strategy should counteract adversary actions, allocate resources, and exploit national strengths.*

16   *This page discusses the need for a competitive space industrial base, reforms in government contracting, and the development of a space commodities exchange. It also suggests the creation of an interagency working group to achieve these goals.*

17   *The page discusses the threats and challenges facing the space industrial base and suggests actions to address them, such as using advanced purchase agreements and a government space development fund, increasing investments in critical space science and technology areas, and exploring public-private partnerships.*

18   *The appendix raises questions about creating a diverse and competitive space industrial base, the vulnerability of the supply chain, the limitations of the Chinese approach, learning from past cases, prioritizing actions, the impact of European space industrial policy, and differentiating between foreign investments.*

19   *The page discusses the role of an import/export bank in supporting the space industrial base and how policymakers should assess finance and insurance costs and access to investment capital for space technology.*

NOTABLE PASSAGES

2   *"The United States of America has no intention of finishing second in space. This effort is expensive — but it pays its way for freedom and for America." - PRESIDENT JOHN F. KENNEDY*

3   *"We are here to underscore the urgency with which all of us must focus our actions to maintain our technological and military dominance. The breadth and depth of Chinese malfeasance with regard not only to our technology, but also to our larger economy and our nation is significant and intentional."*

4   *"Chinese industrial policies of economic aggression, such as investment-driven technology transfer and illegal intellectual property theft, pose a multifaceted threat to our entire national security innovation base."*

5   *"Continued robust development of U.S. civil, commercial and military space activities requires a resilient, domestic, manufacturing and defense industrial base supported by a trusted and responsive supply chain."*

6   *"The accounting firm PwC predicts that China's economy will be more than 40% larger than the United States' by 2050. Timely and proactive measures are critical to disrupting China's plan to become the dominant space power."*

7   *"Refined germanium metal costs have increased 50 percent in the last 30 months and now accounts for nearly 20 percent of the total cost to manufacture finished, completed, satellite solar panels. Manufacturing of satellite solar panels is now impossible without the Chinese controlled raw materials. This provides China with increased influence over the viability of U.S. producers and the availability of radiation hardened solar cells."*

8   *"The unique advantages of space-based capabilities will continue to create a growing commercial, civil and military, space-ecosystem from low Earth orbit (LEO) to geosynchronous orbit (GEO). The satellite architectures within this ecosystem will depart radically from the historic large-satellite-can-do-it-all approach. This ecosystem will be populated with a vastly increased number of assets supporting commercial, civil and military applications across a wide range of satellite sizes, constellations sizes and orbits. The capabilities of these space system architectures will be tailored around power, aperture, bandwidth, interoperability and other functional specifications to maximize network redundancy, efficiency, and value creation."*

9   *"The entire U.S. space industrial base for high-efficiency solar cell production is 20 MOCVD reactors; less than the quantity being installed monthly in China. U.S. producers struggle to consistently fuse that capacity at a level sufficient to maintain its economic viability. The Chinese government's investment of over $10,000,000,000 (ten billion dollars) in MOCVD equipment in San'an Optoelectronics alone has created in less than 10 years an installed base of over 400 MOCVD tools. In such areas where the U. S. space industrial base has become inextricably reliant on its adversaries we can expect them to use their influence in their own national interests."*

10   *"The Nation's interest in assuring economic, political and military superiority in space is best served by a robust national and international space industrial base with a wide range of providers and services. Multiple service providers limit the ability of any one*

provider or nation to exploit market control for economic, political, or military advantage. A diverse set of national and international developers and providers optimizes the effective allocation of resources and drives innovation through competition."

11   "The United States must avoid policies that forcibly shed human know-how and talent, defund essential technology programs, allow foreign acquisition of startup companies or their technologies, or neglect or cede control of key industry and product segments that allow foreign actors unique advantages or critical control over the global space industrial base."

12   "This small responsive launcher will compete head-on with commercial solutions offered in the United States and other western countries. CASIC, one of China's largest state-owned defense contractors, will provide the Kuaizhou-11 through its subsidiary Expace at a highly discounted rate of $5,000 per kilogram to LEO. This is 5 times less expensive than comparable small, responsive launch capabilities. By such predatory, discounting launch rates, Expace threatens the viability of its global competition, which includes the burgeoning launch industry in the US. The same predatory pricing tactic enabled the Chinese drone maker, DJI, to secure 74 percent global market share and decimate the U.S.-based drone industry."

13   "China has a well-understood and effective national strategy to become the global, dominant space power by 2045."

14   "A key vulnerability is the ubiquity of adversary state funding of emerging U.S. space technology innovators. Counterintelligence must continue to work with U.S. companies to block, dilute, and divest adversary funding that provides visibility, influence, and access to these technologies. This must be handled in a nuanced fashion that does not harm revenue, ostracize talent, nor inhibit the speed of innovation. To scale this counterintelligence response effectively requires the support and cooperation of government, industry, and venture capital firms."

15   "U.S. loss as a dominant space competitor would unacceptably impact U.S. civil, economic, political, and national defense power."

16   "include reforms in government contracting and direct government investment as needed to compensate for U.S. adversaries' anti-competitive behavior, and establish the long-term technological and logistical space infrastructure needed to ensure long-term, U.S. dominance in space"

17   "To determine by how much government-wide investments must increase in critical space science and technology areas to support the long-term health of the space industrial base, stimulate the development of key capabilities, and maintain U.S. dominance."

State of the Space Industrial Base:
Threats, Challenges and Actions

A Workshop to Address Challenges and Threats to the U.S. Space
Industrial Base and Space Dominance

Dr. Thomas Cooley, Air Force Research Laboratory

Colonel Eric Felt, Air Force Research Laboratory

Colonel Steven J Butow, Defense Innovation Unit

30 May 2019

The Air Force Research Laboratory's mission is leading the discovery, development, and integration of warfighting technologies for our air, space and cyberspace forces. With its headquarters at Kirtland Air Force Base, N.M., the Space Vehicles Directorate serves as the Air Force's "Center of Excellence" for space research and development. The Directorate develops and transitions space technologies for more effective, more affordable warfighter missions.

The Defense Innovation Unit's mission is to accelerate commercial innovation for national security. It does so by increasing the adoption of commercial technology throughout the military and growing the national security innovation base. DIU's Space Portfolio facilitates Department of Defense partners' ability to access and leverage the growing commercial investment in new space to address existing capability gaps, improve decision making, enable a shared common operating picture with allies, and help preserve the United States' superiority in space.

Cover Photo: Illustration depicting Spaceflight Industries' successful deployment of 64 satellites on its SSO-A dedicated rideshare mission which launched to LEO on 3 Dec 2018 (Credit: Spaceflight Industries).

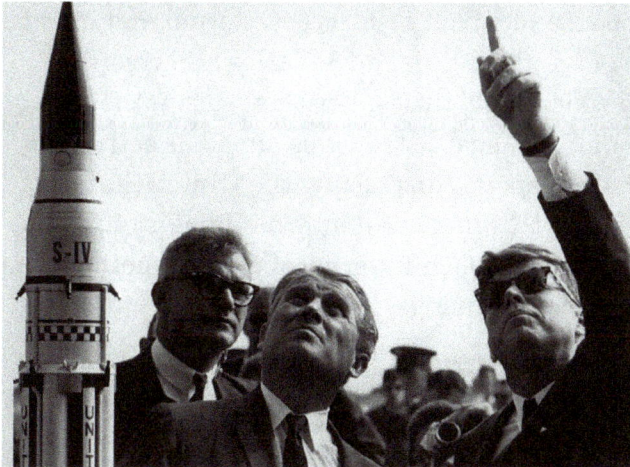

Credit: NASA

The United States of America has no intention of finishing second in space. This effort is expensive — but it pays its way for freedom and for America.

- PRESIDENT JOHN F. KENNEDY

Executive Summary

This white paper is a call to action to address present and emerging challenges and threats to the U.S. space industrial base and space dominance. It summarizes the discussions, conclusions and recommendations from a meeting on March 11 and 12, 2019 of interested parties and experts (see Appendix) gathered from across government, academia and industry sponsored by the Air Force Research Laboratory (AFRL) and the Defense Innovation Unit (DIU). The meeting was motivated by the recognition that in an ever more global and interconnected world, commercial, civil and national defense space capabilities are increasingly vital to national power. To preserve and expand that power requires a coordinated national space strategy and a vibrant, competitive and agile U.S. space industrial base to execute that strategy. The objectives of this meeting were to examine

- the increasing contribution of space to national economic, political and military power;
- the U.S. space industrial base required to ensure and expand that power;
- current and emerging challenges and threats to the space industrial base; and
- potential strategies to address those challenges and threats.

We define the 'U.S. space industrial base' as the private-sector, industry-suppliers of technology, hardware, software, systems, data and financial and insurance capacities that grow the space economy to serve our nation's civilian, civil and national security interests. The U.S. space industrial base is presently a relatively small, nascent part of the national and global economy. As such, it remains particularly vulnerable to, and the government must protect it from, manipulation, distortion, penetration and domination by our adversaries, allies and neutral countries. While the U.S. has long played a dominant role in space and continues to make significant space investments across civil, military and commercial space, the overall domestic effort is insufficiently integrated, focused and leveraged to address the challenges and threats to our Nation's dominant position.

Creating and maintaining the required space industrial base faces external and internal threats and challenges. Externally, our present and potential adversaries and rivals recognize the growing importance of assured space capabilities. For this reason, they have developed and are executing comprehensive national space strategies aimed at actively competing with, complementing, and, in certain respects, displacing, the United States as 'the' or 'one of the' dominant space powers. While this increased international attention poses significant, overarching challenges, China's approach in terms of means, methods and effects presents particular threats to the U.S. space industrial base.

We are here to underscore the urgency with which all of us must focus our actions to maintain our technological and military dominance. The breadth and depth of Chinese malfeasance with regard not only to our technology, but also to our larger economy and our nation is significant and intentional.[1]

The key threatening elements of the Chinese strategy include

- theft of intellectual property combined with a concerted and effective drive to create organic, national expertise across key space science and technology areas;

- direct integration of state-owned corporations and their technologies with commercial, space startup-companies;

- penetration of American companies to obtain and further exploit U.S. technology or to influence those companies in a direction that serves China's domestic space priorities;

- investment in the U.S. space industrial base via front companies and multi-level off-shore accounts to facilitate early venture technology surveillance, infrastructure access and control of developing space capabilities and intellectual property;

- obtaining vertical control of the key space capabilities' supply chains or control of sufficient elements of those supply chains so as to influence space capabilities development in their favor;

- predatory pricing of space capabilities or elements of key space supply chains to control or dominate the market; and

---

1 Testimony of USD(R&E) Mike Griffin to House Armed Services Committee's Military Personnel Subcommittee (June 2018).

- use of state-sponsored venture capital, finance and market control mechanisms to surveille U.S. technology, interdependencies, business model innovations and other advanced concepts.

Chinese industrial policies of economic aggression, such as investment-driven technology transfer and illegal intellectual property theft, pose a multifaceted threat to our entire national security innovation base.[2]

Internally, the challenge is developing an industrial base that outpaces our international adversaries and competitors in speed and innovation in developing new space capabilities and in continually upgrading existing ones. This requires

- upgrade of our own methodologies such as shared, trusted supply chains and interoperable technology standards that accelerate viable commercialization of the space economy;

- the development of more flexible, U.S.-led markets for space capabilities that spread the risk, increase the pool of investors and establishes our Nation's leadership role in setting the international rules for space products and services;

- changes in U.S. government procurement and licensing processes and other regulations to eliminate unnecessary delays and micromanagement of the space industrial base's ability to deliver next generation space capabilities and to enable early U.S. investment in emerging capabilities.

For the United States to be a dominant force in the future space economy during peacetime and to monitor and engage decisively in space when national security is threatened, we require a unified and comprehensive national strategy that builds and continually refreshes a strong space industrial base. The group recommends urgent attention to the development of this strategy as detailed in the Conclusions and Recommendations section of this white paper.

---

[2] Testimony of DASD Eric Chewning to House Armed Services Committee's Military Personnel Subcommittee (June 2018).

Introduction

Continued robust development of U.S. civil, commercial and military space activities requires a resilient, domestic, manufacturing and defense industrial base supported by a trusted and responsive supply chain. In Executive Order 13806, the President directed an assessment of the resiliency of domestic manufacturing, defense industrial base and supply chains[3]. The EO 13806 Report cited numerous cases where domestic production and supply chains had insufficient carrying capacity to surge in response to national emergency or other requirements[4]. The report pointed out U.S. over reliance on China for rare earth minerals and other single points of failure due to reliance on foreign production of unique or commodity products essential to implementing the *2018 National Defense Strategy*[5].

As the nation embarks on reorganizing its space operations and command structure, it is vital to examine how this general problem applies to space[6]. Specifically, this calls for an examination of

- the increasing contribution of space to national economic, political and military power;

- the U.S. space industrial base required to ensure and expand that power;

- current and emerging challenges and threats to the required industrial base; and

- potential strategies to address those challenges and threats.

---

[3] Presidential Executive Order on Assessing and Strengthening the Manufacturing and Defense Industrial Base and Supply Chain Resiliency of the United States (July 21, 2017)

[4] Assessing and Strengthening the Manufacturing and Defense Industrial Base and Supply Chain Resiliency of the United States, *Report to President Donald J. Trump by the Interagency Task Force in Fulfillment of Executive Order 13806* (September 2018)

[5] EO 12806 Report, Figure 14, at page 29

[6] Text of a Memorandum from the President to the Secretary of Defense Regarding the Establishment of the United States Space Command (December 18, 2018)

The Growing Role of Space to National Power

Commercial, civil and military uses of space are rapidly expanding to deliver capabilities and advantages uniquely available from and in space. In the near term, these space capabilities center on information gathering; precision position, navigation and timing (PNT); and broadband communications to include the internet.

For information gathering, no other domain provides equivalent global access. National, commercial, civil and military information dominance is increasingly dependent on space systems' capabilities to observe globally from above, using a rapidly expanding range of sensors refreshed at an ever-increasing time rate, pixel resolution and sensitivity. In an ever more interconnected world, there will be a commensurate or even greater expansion of information flows across the terrestrial, maritime, air and cyber domains. However, in these domains the sources will be localized and prone to greater and easier control, interdiction and corruption by adversaries. Space-based sensors will continue to provide platforms for global observation that are more difficult to disrupt, degrade, and deny than similar sensors in other domains.

The accounting firm PwC predicts that China's economy will be more than 40% larger than the United States' by 2050.[7] Timely and proactive measures are critical to disrupting China's plan to become the dominant space power.

Space will remain the dominant medium for providing precision PNT driven by its global coverage and simplicity of source and applications. The criticality of precision PNT to national infrastructure is evidenced by the continuing proliferation of such space-based systems sponsored by Europe, China, Russia, India, US, Japan, South Korea, and others for civilian, commercial, military and intelligence purposes.

Space communication systems provide global and local capabilities that minimizes supporting ground infrastructure and the need to transmit information on the ground or through the air across the territories of rivals or potential adversaries or areas where the rivals or adversaries could interdict or break the communication path. The recent concern regarding Chinese control of the limited number of fiber cable connections is a case in point. In addition, space communication systems can achieve higher latency than ground-based, global, fiber systems and equivalent bandwidth to existing ground communication networks through laser cross-, up- and down-links.

---

[7] Feffer, J. (2019). The Widening Rift Between the U.S. and China. The Nation.

Figure 1: Germanium wafers manufactured from the National Defense Stockpile (Credit: DLA).

Germanium Wafer Production

Refined germanium wafers are the basis for nearly 100 percent of the high-efficiency, radiation hardened solar cells that power satellites today. China accounts for over 70 percent of the world's germanium mining, refining and production while the United States contributes nothing to mining and only ~2.5 percent to the world's refined germanium output[8]. China has safeguarded its market pricing dominance by purchasing and storing germanium and by increased export tariffs[9]. Refined germanium metal costs have increased 50 percent in the last 30 months and now accounts for nearly 20 percent of the total cost to manufacture finished, completed, satellite solar panels. Manufacturing of satellite solar panels is now impossible without the Chinese controlled raw materials. This provides China with increased influence over the viability of U.S. producers and the availability of radiation hardened solar cells.

---

[8] U.S. Department of the Interior, U.S. Geological Survey, Mineral Commodity Summaries, pp. 70-71, 2017.
[9] Global and China Germanium Industry Report, 2013-2016, ResearchInChina, 2014.

The unique advantages of space-based capabilities will continue to create a growing commercial, civil and military, space-ecosystem from low Earth orbit (LEO) to geosynchronous orbit (GEO). The satellite architectures within this ecosystem will depart radically from the historic large-satellite-can-do-it-all approach. This ecosystem will be populated with a vastly increased number of assets supporting commercial, civil and military applications across a wide range of satellite sizes, constellations sizes and orbits. The capabilities of these space system architectures will be tailored around power, aperture, bandwidth, interoperability and other functional specifications to maximize network redundancy, efficiency, and value creation. Within this ecosystem, space broadband communications and internet capabilities will move from a small number of large GEO satellites to a mixed architecture of large GEO satellites and proliferated constellations of large numbers of small satellites at lower orbits. We can also expect first sub-orbital, and orbital space tourism to become a part of this ecosystem.

As in other domains, the commercial space industrial base will need to provide end-to-end delivery of a significant portion of critical civil and military capabilities, such as communication bandwidth, imagery, launch, debris removal and other commoditized services. There will be an increasingly symbiotic relationship between the economic development of LEO and GEO space and increased military, civil, commercial and intelligence surveillance and reconnaissance of actors and their activities in LEO and GEO space with commercial systems both being assets to be monitored and sources for monitoring information, when appropriate.

In the mid- to long-term (5 years and beyond), the development and deployment of systems and capabilities beyond the LEO and GEO ecosystem, will have two drivers: first, by the military's need to expand the locations and operations of critical assets into cislunar space to limit adversaries' abilities to detect and attack these assets and to enhance ours and our adversaries' ability to apply force through, from and in space; and second, it will be driven by the need to establish the required infrastructure and capabilities to return and then establish a permanent U.S. presence on the Moon and beyond.

The resulting technology, infrastructure and capabilities will establish the foundations (including supply chain logistics) for the extension throughout the cislunar domain of military power and for the economic exploitation through space manufacturing, space power and resource extraction. The foundation for a sustainable space economy, such as cislunar infrastructure, strategically depends on close collaboration with national commercial capabilities and the maintenance of a strong space industrial base. Such an approach maximizes the U.S. position to lead in the economic exploitation of space.

Figure 2: Metal Organic Chemical Vapor Deposition (MOCVD) reactor (Credit: Veeco).

High Efficiency Solar Cell Production

The annual worldwide market for high-efficiency solar cells and panels for space applications has been approximately 750,000 Watts.[10] With industry plans for mega-constellations, forecasted demand for space solar power exceeds 2,000,000 Watts annually. A minimum of a 5X growth in the next 5 years in solar cell and panel production will be required to support recent FCC and ITU filings for mega-LEO constellations. These high-efficiency, radiation hardened solar cells for spacecraft are made using a Metal Organic Chemical Vapor Deposition (MOCVD) reactor. China became the world leader in MOCVD capacity in 2012, with an installed base surpassing 1,000 tools.[11] In 2018, 330 such tools were installed for Gallium nitride (GaN) production alone.[12] The entire U.S. space industrial base for high-efficiency solar cell production is 20 MOCVD reactors; less than the quantity being installed monthly in China. U.S. producers struggle to consistently fuse that capacity at a level sufficient to maintain its economic viability. The Chinese government's investment of over $10,000,000,000 (ten billion dollars) in MOCVD equipment in San'an Optoelectronics alone has created in less than 10 years an installed base of over 400 MOCVD tools. In such areas where the U. S. space industrial base has become inextricably reliant on its adversaries we can expect them to use their influence in their own national interests.

---

10 Semiconductor Today, Compounds & Advanced Silicon, Vol. 5, Issue 7, pp. 94-98, 2010.
11 LEDs Magazine, 2011.
12 Market Insight, IHS Markit, 2018.

Major Findings

We present our findings in two parts: first, findings about the role of the U.S. space industrial base in sustained space superiority, and second, findings on the actions of China and other rivals and adversaries that threaten and challenge the U.S. space industrial base and space power.

General Findings on the Criticality of Space and the Space Industrial Base

- The Nation's interest in assuring economic, political and military superiority in space is best served by a robust national and international space industrial base with a wide range of providers and services. Multiple service providers limit the ability of any one provider or nation to exploit market control for economic, political, or military advantage. A diverse set of national and international developers and providers optimizes the effective allocation of resources and drives innovation through competition.

- Military mission performance and mission assurance are significantly enhanced by a strong national and international industrial ecosystem. The range of options for services in such an ecosystem improves the resilience of military capabilities through increasing the diversity, distribution, disaggregation, deception and proliferation of means by which critical space-based warfighting capabilities are provided. Such an ecosystem strengthens the defense of space systems and capabilities by greatly increasing the number and complexity of space assets that an adversary must disrupt or destroy to deny capabilities in time of conflict. It strengthens mission assurance through enabling diverse means to reconstitute military space capabilities in case of loss or degradation due to attack.

- A whole-of-government, national space strategy to include a tailored space industrial policy is required to address the challenges to U.S. space interests. This strategy must promote the type of national and global market in which American business excels -- a dynamic, open and competitive, commercial market that drives the efficient application of capital and resources, and maximizes on-going innovation through competition. Specifically, the development of a space capabilities commodity exchange is required to create greater distribution of risk, a greater range of investment sources, and to provide a rules-based system limiting any nation's ability to pursue anti-competitive strategies and practices. This must include the investments across government, industry and academia to maintain space science and technological advantage over our adversaries.

- Sustainment and further development of the U.S. space industrial base is impeded by its present, over-reliance on the government as the primary consumer of space products and capabilities. The requirements-driven, lengthy, and risk-averse procurement process for government capabilities is poorly suited to the rapid, learning-by-doing development of the commercially-oriented, internationally competitive, space industrial base. Government acquisition approaches for space capabilities must adapt to take advantage of the commercially-oriented, space industrial base and promote its development and expansion. Government development and purchase of space capabilities must take a longer-term view of the investments needed now to create the industrial base of the future, understanding that these investments may entail risk (e.g. long-term commitments to buy commercial products early in the commercial development process).

- The United States can take immediate action to address adversarial activity that directly challenges U.S. space dominance. This includes increased focus on and commitment of resources toward combating corporate- and state-sponsored espionage. The United States must avoid policies that forcibly shed human know-how and talent, defund essential technology programs, allow foreign acquisition of startup companies or their technologies, or neglect or cede control of key industry and product segments that allow foreign actors unique advantages or critical control over the global space industrial base.

- The future and growth of the space economy is critically dependent on continuing reductions in the costs and risks associated with launch. There is a bifurcation of launch providers between lower-cost, "bulk" carriers (SpaceX, ULA, Blue Origin, etc.), and higher-cost, "niche" providers offering lower lift-mass but launch to a specific orbit at higher costs. To date, the United States has been the driver behind increased interest in launch innovation but foreign government-supported launch programs are a serious threat to the development and maintenance of a robust, competitive, open market.

Figure 3: Kuaizhou-1A and 11 launch vehicle models on display. (Credit: SpaceNews).

State-Sponsored Commercial Launch Services

The China Aerospace Science and Industry Corporation's (CASIC) Kuaizhou-1A, solid-propellant, mobile, launch vehicle is the baseline for the new, "commercial" Kuaizhou-11 launch system. This small responsive launcher will compete head-on with commercial solutions offered in the United States and other western countries. CASIC, one of China's largest state-owned defense contractors, will provide the Kuaizhou-11 through its subsidiary Expace at a highly discounted rate of $5,000 per kilogram to LEO.[13] This is 5 times less expensive than comparable small, responsive launch capabilities. By such predatory, discounting launch rates, Expace threatens the viability of its global competition, which includes the burgeoning launch industry in the US. The same predatory pricing tactic enabled the Chinese drone maker, DJI, to secure 74 percent global market share and decimate the U.S.-based drone industry.[14]

---

[13] Jones, A. (2019). Chinese state-owned firms preparing to launch new commercial rockets. SpaceNews.
[14] Bateman, J. (2017). China drone maker DJI: Alone atop the unmanned skies. CNBC.

- There is increasing evidence that much of space-based communication and internet infrastructure will move from GEO to LEO driving the development and deployment of large constellations of small satellites (hundreds to thousands). This will not only expand the satellite bus and payload market but also, on a global scale, will catalyze major growth of the space industrial base. The country that is best positioned to support and exploit such a move will have a distinct advantage as a dominant space power.

General Finding on the Challenges of Coordinated Chinese and other Rival Space Industrial and Military Policies and Actions

- China and other adversaries/rivals recognize the growing economic, political and military potential of space.

  o Economically, they recognize that space is becoming a key element of the global infrastructure in the near term for communications, intelligence, surveillance, and reconnaissance, and PNT and, in the long term, as a source for tourism, power generation, space manufacturing, and resource extraction.

  o Politically, they recognize the advantage of space in the collection and control of data globally, the rapid conversion of the data into information, and the speedy transmission of that data globally to drive political decision making.

China has a well-understood and effective national strategy to become the global, dominant space power by 2045.[15]

---

  o Militarily, they recognize that space provides distinct advantages in global awareness of the dynamic battlefield of the future, in converting space-obtained data into critical information, in ensuring continuous location of friendly and adversary forces, and in rapidly and securely transmitting globally information from space and the other domains to drive timely decision making and the appropriate and efficient application of force.

- China has a well-understood and effective national strategy to become the global, dominant space power. This is reflected in the Spatial Information Corridor component of China's Belt and Road Initiative. A key part of this strategy is to penetrate and dominate elements of the global space industrial-base while

---

[15] Erwin, S. (2019). Congressional panel looks at national security implications of China's space ambitions. Space News.

developing their own strong national space industrial base. This strategy includes a whole-of-government, integrated plan to

o   develop their indigenous space industrial base;

o   steal space intellectual property;

o   develop organic expertise in key areas of space science and technology through association with the United States' and its allies' universities, research centers and startups;

o   penetrate global space companies by strategic investment, human intelligence, cyber intrusion, or other means to control or leverage their proprietary capabilities to support Chinese operational and technical dominance;

o   exercise vertical or significant horizontal control over the supply chain to influence or control the development of critical capabilities and determine winners and losers within the global space industrial base;

o   penetrate foreign space companies to provide access to space-enabled global infrastructure; and

o   use predatory pricing and unfair trade practices to dominate key market segments.

- A key vulnerability is the ubiquity of adversary state funding of emerging U.S. space technology innovators. Counterintelligence must continue to work with U.S. companies to block, dilute, and divest adversary funding that provides visibility, influence, and access to these technologies. This must be handled in a nuanced fashion that does not harm revenue, ostracize talent, nor inhibit the speed of innovation. To scale this counterintelligence response effectively requires the support and cooperation of government, industry, and venture capital firms.

- Legal changes are required to stop the hemorrhaging of technology. The most valuable technologies are often emerging and thus often not yet protected by copyright or patents. This poses an enormous vulnerability that is exploited by our adversaries. The Department of Justice must develop the tools to protect intellectual property during early phases of technology development.

- The dominant position that Huawei has created in 5G communications is an example of the type of challenge that the Chinese space strategy poses for the nation and the U.S. space industrial base.

- China's industrial base has already positioned itself to potentially dominate the mass market for space-based solar-power systems for planned proliferated LEO constellations.

Conclusions and Recommendations

- U.S. loss as a dominant space competitor would unacceptably impact U.S. civil, economic, political, and national defense power.

- The challenge and threat of the Chinese and other adversaries and competitors to the U.S. space industrial base is significant and growing. This must be addressed now.

- Competing in the space economy and securing national security interests are best achieved using agile and adaptive strategies that best combine and leverage the roles of government, industry and academia.

What is required is a whole-of-government, national space strategy to include a tailored space industrial policy.

- A whole-of-government plan must be urgently developed that

  o recognizes the adverse actions of our adversaries;

  o develops the upgraded policies and actions to counteract them; and

  o allocates the leadership, institutional and other resources to implement them.

- A national strategy is required that includes a space industrial base policy. It must

  o be whole-of-government in order to be easily navigated by the private sector and readily administered inside of government for rapid decision-making and analysis;

  o exploit national strengths, and not become reactive or imitate the bad behavior of our rivals and adversaries;

- o create a nationally and internationally competitive space industrial base and development environment that allow the efficient allocation of resources and maximizes innovation;

- o include reforms in government contracting and direct government investment as needed to compensate for U.S. adversaries' anti-competitive behavior, and establish the long-term technological and logistical space infrastructure needed to ensure long-term, U.S. dominance in space;

- o include an integrated, national approach to ensure U.S. parity (if not superiority) in the science and technology driving the development of new space capabilities;

- o include the development of a space commodities exchange to spread risk, add liquidity, increase the diversity and capital of space investors, and drive international standards for the development and sale of interoperable space capabilities and marketable financial and risk-transfer (insurance) instruments;

- o include an Export Bank or similar financial support options for developing countries to finance goods and services as a means of aiding American companies in the emerging global space market;

- o be a strategy where the United States plays a lead role in developing norms of state behavior and a global space industrial base that minimizes our rivals' and adversaries' ability to dominate the market by leveraging non-market strategies or ambiguities in international space law; and

- o inspire, educate and grow the human capital necessary to sustain a strong and prosperous space industrial base for the future.

- • The National Space Council should immediately institute an interagency and commercial working group to provide an actionable plan in six months to achieve the following goals

  - o create a U.S.-based space commodities market;

  - o reform government space acquisition processes to increase the use of innovative, simplified, competitively awarded agreements to reduce timelines and costs of procuring new capabilities (e.g.: explore setting aside

20 percent of space total obligation authority (TOA) for acquisition of nontraditional commercial capabilities)[16];

- o use advanced purchase agreements, when feasible, as a means of more effectively signaling government interest in procuring innovative goods or services and incentivizing the private capital market to provide for their funding and development;

- o use a government space development fund, when feasible, to assist emerging space industrial base companies competitively disadvantaged due to Chinese actions and viewed vital to the long-term viability and leadership of the U.S. space industrial base;

- o identify the most detrimental actions by China and other adversaries and competitors, and determine the required response to mitigate and deter them in the future;

- o determine by how much government-wide investments must increase in critical space science and technology areas to support the long-term health of the space industrial base, stimulate the development of key capabilities, and maintain U.S. dominance; and

- o explore public-private partnerships to invest in space infrastructure that would ensure U.S. leadership in the growth of future space-enabled economic engines such as smart cities, connected logistics, precision agriculture, autonomous vehicles, and others.

---

[16] Total obligation authority (TOA) is a financial term expressing the value of the direct Defense program for a fiscal year, exclusive of the obligation authority from other sources such as reimbursable orders accepted.

Appendix

Where do we need more information?

- What are all the elements to be pursued to create a diverse, competitive and innovative domestic space industrial base? What in terms of creating the desired market is required besides establishing a space commodities market?

- Better determination of the fragility of space capabilities to market manipulation or domination? If the cost of entry is low and the return on investment high, then market domination is hard. Building aircraft versus operating an airline.

- How vulnerable is the supply chain to domination? How can the space economy grow in a manner that reduces supply chain vulnerability in each growth phase? In general, space companies will seek to avoid dependence on single suppliers for critical elements of the capabilities they produce. They would prefer standardized, interoperable elements to aid in lowering cost and diversifying sources of supply.

- What are the longer-term limitations to the Chinese approach? Historically, state centrally-controlled systems are inefficient in the application of capital and in technological innovation and refresh. Eventually the inefficient application of resources (both fiscal and intellectual) has negative repercussions. As China uses its economy to project soft power, how should such power's influence be met?

- How different is the Chinese approach from what other developing countries pursued in the past (the US?)? What can we learn from these cases?

- What parts of the problem should be addressed in what order? For example, should the greater focus be on promoting the U.S. industrial base or in preventing intellectual property (IP) leakage to the adversaries and their penetration of U.S. companies? Which is more important: plugging holes or stimulating the rate of change? Ultimately, resources are limited so it is critical to decide what to do, in what order and with what allocation of resources applied to the parts of the problem addressed.

- Quantify the degree to which European space industrial policy affects present and future of U.S. space industrial base. What can we learn both positively and negatively from the European approach?

- How do we differentiate between investment by foreign entities for the purpose of investment and for purposes inimical to U.S. national interests?

- What role can or should an import/export bank play in supporting the development of a strong, national industrial base? As a general matter of space trade and development policy, how should U.S. public and private policymakers quantify alternative features of national schema for finance and insurance costs and access to investment capital willing to take risks of space technology?